Juvénal BISIMWA MUGOBE
John Peter BISIMWA RUGUSHA

EFFECTS OF ARMED CONFLICT ON PROTECTED AREAS IN DRC

Juvénal BISIMWA MUGOBE
John Peter BISIMWA RUGUSHA

EFFECTS OF ARMED CONFLICT ON PROTECTED AREAS IN DRC

CASE OF KAHUZI BIEGA

ScienciaScripts

Imprint

Any brand names and product names mentioned in this book are subject to trademark, brand or patent protection and are trademarks or registered trademarks of their respective holders. The use of brand names, product names, common names, trade names, product descriptions etc. even without a particular marking in this work is in no way to be construed to mean that such names may be regarded as unrestricted in respect of trademark and brand protection legislation and could thus be used by anyone.

Cover image: www.ingimage.com

This book is a translation from the original published under ISBN 978-620-3-44689-0.

Publisher:
Sciencia Scripts
is a trademark of
Dodo Books Indian Ocean Ltd. and OmniScriptum S.R.L publishing group

120 High Road, East Finchley, London, N2 9ED, United Kingdom
Str. Armeneasca 28/1, office 1, Chisinau MD-2012, Republic of Moldova, Europe
Printed at: see last page
ISBN: 978-620-5-72543-6

EFFECTS OF ARMED CONFLICT ON PROTECTED AREAS IN THE DRC: THE CASE OF

KAHUZI BIEGA

With the assistance of Mr.

CT Juvénal BISIMWA MUGOBE
UDDAC/BAGIRA in Bukavu, DRC.
juvenalmugobeb@gmail.com
Telephone: (+243) 825 255 354

John Peter BISIMWA RUGUSHA
Environmental Management Technician and Air
Station Manager in Lulingu/Tchonka, DRC Phone:
(+243) 824 220 586

Juvénal BISIMWA MUGOBE, born on March 08, 1974 in Murama, in the Locality of Murama, Grouping of Bushumba, Territory of Kabare, Province of South-Kivu in the City of Bukavu in the Democratic Republic of Congo. Head of works at the University of Sustainable Development in Central Africa.

December 2022

LIST OF ACRONYMS

ANEE: National Association for Environmental Assessment

PA: Protected Areas

ICRC: International Committee of the Red Cross

COCO: Community Conservation

ENMOD: Environmental Modifications

FARDC: Armed Forces of the Democratic Republic of Congo

FDLR: Democratic Forces for the Liberation of Rwanda

GRIP: Research and Information Group on Peace and Security

GTCAR: Armed Conflict Working Group

ICCN: Congolese Institute for Nature Conservation

ISTD: Higher Institute of Development Techniques

MONUSCO: United Nations Mission for the Stabilization of the Congo

WHO: World Health Organization

NGO: Non-Governmental Organization

NTFP: Non-Timber Forest Products

PNKB: Kahuzi-Biega National Park

UNDP: United Nations Development Programme

UNEP: United Nations Environment Programme

PNVi: Virunga National Park

RM: Raia Mutomboki

NR: Natural Resources

Sd: No date

SEEAC: Sub-regional Secretariat for Environmental Assessment in Central Africa

TFC: End of Cycle Work

IUCN: International Union for Conservation of Nature

UNESCO: United Nations Educational, Scientific and Cultural Organization

WCPA: World Commission on Protected Areas

%: Percentage

SUMMARY

The overall aim of this work was to contribute to the protection of the PNKB and its natural resources at all times, even during periods of armed conflict, in order to maintain its overall ecological, socio-economic and even cultural value.

After the field research, we obtained several results of which below are the major ones:

100% of respondents recognized the presence of armed conflict actors or armed groups in and around the KBNP in the Lulingu sector, notably the FARDC and the MR (60.7% of respondents) and the basis for the presence of armed groups in the KBNP in the Lulingu sector, which are the weakness in the restoration of state authority in the area (60.7% of respondents), the ignorance of the importance of the park by these armed groups (23.8% of respondents) and finally the weakness of the park surveillance service (15.3% of respondents);

Three types of ecological effects on PNKB linked to the presence of these armed groups in and around the park were recognized by the respondents, namely the continued loss of species of fauna and flora and other NR in the park (51.5% of respondents), the loss of habitat for the park's fauna and flora (33% of respondents), and finally the pollution of the park's ecosystem (15,5% of respondents), which directly associate three types of socio-economic effects, namely the loss of revenue from tourist activities (42.3% of respondents), the loss of benefits initiated by the park for the local population (39.2% of respondents), and finally the loss of work for park agents (18.4% of respondents).

Key words: *Armed conflicts, Effects, Protected areas, PNKB, Lulingu.*

CHAPTER ONE
GENERAL INTRODUCTION

1.1. Status of the issue

In order to deal with our research topic, we first consulted the works of certain researchers who preceded us and who spoke about the effects of armed conflicts on the environment in general and on PAs in particular, either in DR Congo or elsewhere in the world. Among these researchers we can cite :

1) AL-HAMANDOU and MICHEL-ANDRE (2016) in their article entitled "Armed Conflicts and the Environment" the authors point out that beyond the obvious humanitarian issues, armed conflicts raise important and crucial environmental issues. For these authors, armed conflicts are also accompanied by a collapse of environmental governance, which in turn leads to accelerated environmental degradation.

   Sometimes destruction causes irreversible degradation in ecosystems; this is the case when species may be driven to extinction, or fragile ecosystems may be irreversibly degraded, or resources irreversibly destroyed or contaminated. All institutional protection systems, such as Protected Areas or National Parks, become catchment areas for displaced people or combatants, with immediate, and often irreversible, consequences for the quality of these ecosystems.

   These authors show that the environment is both an aggravating factor and a cause of armed conflict and that an environmental assessment in armed conflict situations is indispensable.

2) MUPILI (2012) in his Master 2 thesis at the University of Limoges in environmental law conducted in the DRC entitled "Problems of the application of international environmental law in the fight against violations of environmental law by armed groups in eastern DRC", The author shows that the DRC has a high natural potential throughout its territory, however, for several years now the ecological situation has been deteriorating day by day in the east of the DRC following a succession of wars with no hope of real pacification despite the presence of MONUSCO forces alongside the FARDC.

The changing nature of the war has taken many forms. In addition to these successive conflicts, there is a permanent armed conflict in the east of the country, where the actors are armed militias that escape the control of the state authority and therefore exercise their powers without observing either humanitarian law or environmental law. They set up their headquarters and confine their fighting to the forests in general and to the PAs in particular, where they damage the ecosystem.

For this author, the actors in armed conflicts, particularly armed groups, set up their headquarters in protected areas where they live by hunting protected animals in the parks, they cut down trees for their homes and to make firewood, illegally exploit mines without an environmental and social impact study in the rivers and mountains, and finally the pollution of water, the source of several contagious and infectious diseases leading to fatal epidemics. The author concludes by saying that the damage caused to these resources can lead, long after the conflicts, to harmful and even lethal effects on the affected populations.

3) The GTCAR (2004) in collaboration with ANEE and SEEAC, during a Workshop under the High Patronage of His Excellency the Minister of Environment, Conservation of Nature, Water and Forests of the DRC on "The environmental stakes and impacts of armed conflicts in the DRC", which concerned the environmental aspects and stakes in the context of armed conflicts in the specific theaters of North Kivu, South Kivu and the Oriental Province of the DRC told the workshop participants that the armed conflicts that the DRC has experienced in recent years have had, beyond the catastrophic humanitarian consequences, direct and indirect consequences on the environment and that this aspect of the conflicts is often neglected because it is not a priority in conflict situations, yet it is critical, because environmental damage can irreparably compromise the chances of rehabilitation and reconstruction.

Among the priority objectives of the workshop was to become aware of the impacts and environmental issues of different nature related to armed conflicts, in general, and those that took place in the DRC in particular, whose examples focused on the PAs, including the Virunga Park, which is under threat from the gorilla population and the destruction of the bamboo forest by armed groups

4) PALUKU (2005) in his study entitled "Effectivité de la protection de la biodiversité forestière en République Démocratique du Congo: Cas du PNVi" conducted in the DRC, precisely in the province of North Kivu in the PNVi, shows that the DR Congo

has not escaped the misdeeds of war and the biodiversity of the PNVi has reached a very low level from the quantitative and qualitative point of view. War and economic exploitation have always been closely linked.

For this author, deploring the harms of armed conflict, James and Oglelhorpe write: the destruction of habitat and the disappearance of wildlife resulting from war are among the most widespread effects; as a result of the destruction of their habitat, some species of fauna and flora may be threatened with local extinction or even disappear. Political instability during wartime often has the immediate consequence of making it impossible for residents to grow crops.

To survive, they are gradually forced to turn to wild foods. In areas where fighting is taking place, troops regularly hunt large numbers of mammals for food. In times of armed conflict, those in power often feel an urgent need for income. In order to finance their military activities, they may turn to the extraction of natural resources such as timber and ivory for commercial purposes.

5) According to EMMANUELLE (2014), since 1945, the year marking the end of World War II, the number of armed conflicts has gradually increased to reach a peak in the early 1990s. For the author, between 1946 and 2009, there were just under 250 armed conflicts in 141 different places. Whatever the motivations behind these wars and their magnitude, they have at least one common denominator: the human drama created by their consequences.

Indeed, the impacts on civilians, whether collateral damage or carefully planned military tactics, generate or exacerbate real humanitarian crises whose traces may persist long after the war. This author concludes by saying that apart from the effects on civilian populations, armed conflicts have numerous impacts on the components of the natural environment, including biological diversity. These impacts are mostly negative and are a direct or indirect result of the conflict.

1.2. Issue

To preserve the ecological balance of natural ecosystems, whose signs of degradation were already visible since the 19th century[e] , one of the first measures that were imposed on a global scale is that proposed by the IUCN in 1994. It aims at the creation of protected areas and the reinforcement of protection measures of those already existing (WHITE and EDWARDS, 2001; RAMADE, 2002; KATOTO, 2012).

At the global level, protected areas are considered as the guarantors of the preservation of a collective heritage still heavily impacted by the development of our civilization. The fauna, flora and habitats that they make it possible to safeguard are of universal value not only as original species and spaces, but also because of the current or potential role that they can play for mankind. It is therefore necessary for everyone to feel concerned by the management of any protected area, on any surface of the globe (PATRICK, 2009).

In Africa, protected areas cover a wide range of ecosystems, from coral reefs to mangroves, equatorial forests and deserts. Some countries, such as Cameroon, offer a great diversity of ecosystems, allowing, in a few hundred kilometers, to go from Sahelian environments to the most impenetrable rainforest. This wealth of habitats allows for the existence of a wide variety of flora and fauna, which figure prominently in the imagination of children in many countries. On this continent, part of human life and activities revolve around nature (PATRICK, 2009).

However, in most of these African PAs, habitat destruction and wildlife loss are among the most widespread and severe effects caused by armed conflict, which may be strategic, commercial, or subsistence in origin.

For example, vegetation may be cut, burned, or defoliated to increase mobility and visibility of troops. In Rwanda in 1991, the army mowed a 50-100 m wide strip through the bamboo forest adjacent to the Virunga volcanoes in order to reduce the risk of ambush along a major trail (KALPERS, 2001). In areas where fighting is taking place, troops regularly hunt large numbers of large mammals for food.

This practice can have disastrous consequences for wildlife populations, especially if military activities continue over a long period of time. When displaced people are temporarily relocated, they cut vegetation for agricultural purposes or to obtain firewood. Such practices quickly lead to deforestation and erosion. Since refugees and displaced persons are often resettled in marginal and vulnerable ecological areas, environmental recovery is limited.

The political instability that prevails during wartime often has the immediate consequence of making it impossible for residents to grow crops. To survive, they are gradually forced to turn to wild foods such as meat and food plants. At the same time, displaced people collect firewood, food plants and other local natural resources.

In Central Africa, particularly the Congo Basin, known for its unique biodiversity, there are a significant number of PAs. However, violent armed conflicts continue to abound in this region of the world. The ravages of war on human populations and the development of countries that suffer from it are no longer to be proven. It is now well established that direct and indirect impacts of armed conflict also occur on the natural environment and biological diversity (EMMANNUELLE, 2014).

In the DRC, no one is unaware that with a surface area of 2,345,409 km^2 this country abounds in several natural resources, including A diversified fauna including 409 species of mammals that is 54.1% of the species inventoried in Africa, 1,086 species of birds of which 655 are endemic to the DRC. The Congolese flora as a whole, all phyla combined, includes nearly 377 families, 2,196 genera and 10,324 species. An important and dense hydrographic network that covers almost 86,080 km^2 or (3.5%) of the country's surface.

Thus, the country proceeded to the creation of national parks, the first of which dates back to 1925. Currently, there are seven national parks, with a total area of 8,240,000 ha (PALUKU, 2005). However, conflicts have led and continue to lead to a series of consequences on the ecosystems by disrupting the ecological balance. These consequences are manifested both in the forests and in the protected areas. During this period, the phenomenon of poaching, the uncontrolled cutting of wood and the exploitation of other resources in PAs has developed. Armed conflicts have caused a movement of the population from villages to the forests.

The ecosystems of eastern DRC have been strained from the onset of armed conflict to the present day (EMMANNUELLE, 2014). PNKB is one of the national parks established in the eastern part of the country. It was created in 1937 by the colonial authority under the name "Réserve intégrale Zoologique et Forestière de Kahuzi-Biega", covering an area of 75,000 ha and is to help protect the habitat of Grauer's gorilla (*Gorilla gorilla graueri*).

In 1970 the reserve obtained the status of a national park through the revision of part of the boundaries of the southern part. The area was reduced from 75,000 ha to 60,000 ha. In 1975, the inclusion of the lowland forests, connected to the mountainous area by a corridor, increased the area to 600,000 ha. The objective of this new modification was to protect a continuum of high and low altitude forest habitats and to ensure the exchange of mammal populations between the two forest blocks, in particular gorillas and elephants.

However, the security situation remains precarious both within and at the edge of this PA, as the region is the scene of armed conflicts between different armed groups involving the loyalist force on the one hand and between different local militias on the other, not to mention the aftermath of elements of the Rwandan rebel movement FDLR that established itself in this park several years ago (UWE and TERESE, 2006). All sectors of the park have experienced the presence of armed groups, whether FARDC or local militias of any kind. The Lulingu sector is not spared from this situation.

The presence of armed groups, but also of the FARDC in this part of the park, is not without effects on its conservation and on the maintenance of the ecological balance in general in this protected ecosystem. From this analysis, we are obliged to ask ourselves the following questions:

Why are the actors of the armed conflicts in the PNKB in the Lulingu area?

What are the effects of the presence of armed conflict actors on the KBNP in the Lulingu area in order of importance?

How can the presence of armed conflict actors in and around the Lulingu area of the KBNP be addressed?

1.3. Assumptions

According to RONGERE (1971), "the hypothesis is the proposal of answers to the questions one asks about the research object, formulated in such terms that observation and analysis can provide an answer, it is nothing more than the guiding thread for a committed researcher." To this end, we propose the following provisional answers:

Actors of armed conflict are said to be in the PNKB in the Lulingu area for various reasons;

The effects of the presence of armed conflict actors on the KBNP in the Lulingu area are said to be diverse and of an ecological and socio-economic nature;

It would have a number of features that could help address the presence of armed conflict actors in and around the KBNP in the Lulingu area.

1.4. Objectives of the work

The overall objective of this work is to contribute to the protection of the PNKB and its natural resources at all times, even during periods of armed conflict, in order to maintain the overall ecological, socio-economic and even cultural value of the Park.

The specific objectives of this study are:

Identify the reasons for the presence of armed conflict actors in and around the KBNP in the Lulingu area;

List the ecological and socio-economic effects of the presence of armed conflict actors on the KBNP in the Lulingu area in order of importance;

Propose mechanisms that can contribute to the fight against the presence of armed conflict actors in and around the PNKB in the Lulingu sector.

1.5. Choice and interest of the subject

In any case, the environmental issue is deferred to the humanitarian issue. However, after armed conflicts, the environment and its resources should be the basis for reconstruction. We know the importance of water, biodiversity, forests and agricultural areas. The damage caused to these resources can have harmful, even lethal, effects on the affected populations long after the conflict.

It is essential on the one hand to demonstrate the critical state of the environment due to these recurrent armed hostilities and on the other hand to warn the actors of the armed conflicts that these acts have serious repercussions on the ecosystems in general and the PAs in particular in terms of degradation of their natural resources that can affect the very existence of the human being

This study has a theoretical contribution to any reader and researcher to be informed on the fallout of armed conflicts, on the PAs and their natural resources and to challenge the State in front of its responsibilities in order to fight against the proliferation of armed groups in and around the PAs in the concern of avoiding the destruction of ecosystems by perfectly supervising the actors of these armed conflicts.

On a practical level, this study helps to address the deficit of PA natural resource conservation during armed conflict.

On a personal level, this research helps us to improve our knowledge of the impacts of human activities on PAs in general and armed conflict in particular.

1.6. Work delimitation

This work is determined in time and space.

Spatially, this study focuses on the KBNP in the Lulingu sector, Shabunda Territory, South Kivu Province, DRC.

From a temporal point of view, our investments cover the period from 1998 to 2016.

1.7. Labor Subdivision

This work, apart from the introduction, suggestions and conclusion, will be the subject of three chapters that follow:

> The first chapter presents the general introduction;

> The second chapter gives the literature on the subject;

> The third chapter discusses the survey, analysis and interpretation of the results.

1.8. Difficulties encountered

This study was not without its difficulties. The field research was not easy because it took place in an area that is not only landlocked but also home to a multitude of uncontrolled armed groups, the RMs. We therefore spent a lot of money to access information related to our subject of study. Nevertheless, we have used some strategies to bypass them and to be able to present this work to you today.

CHAPTER TWO

REVIEW OF THE LITERATURE ON THE SUBJECT

2.1. Understanding of the key concepts of the subject

2.1.1. STUDY

2.1.2. EFFECTS

2.1.3. ARMED CONFLICTS

Etymologically, the term conflict comes from the Latin "*conflictus*" which means clash, fight, struggle, antagonism of interests or powers. It is also defined as a process involving reactions and behaviors that begin when one party perceives that it has been or will be insulted by another party. Thus, conflicts are the result of the search for contrary or apparently incompatible interests between individuals, groups or countries (CHARLINE and CHARLES, Sd).

For this reason, the concept of armed conflict applies to different types of confrontations that can occur between two or more state entities, between a state entity and a non-state entity, between a state entity and a dissident faction and/or between two ethnic groups within a state entity using firearms or knives (VERRI, 1988).

2.1.4. PROTECTED AREAS (PA)

According to the updated definition of the IUCN (2008) a protected area is "a clearly defined geographical space, recognized, dedicated and managed, by any effective legal or other means, to ensure the long-term conservation of nature and its associated ecosystem services and cultural values". This simple, concise definition determines the fundamental objectives of protected areas: protection and maintenance of biodiversity (understood in its three dimensions: genetic, specific and ecosystemic), natural resources, landscapes and related cultural values.

For IUCN, only areas whose primary purpose is to conserve nature can be considered protected areas. Protected areas may have other equally important purposes, but in the event of a conflict of interest, nature conservation must take priority. Man is not excluded from protected areas and their management, but has a rightful place in them without becoming a factor of impoverishment, pollution, disturbance or trampling. The task of the manager is outlined; he must constantly play on a subtle balance between the sometimes contradictory interests arising from the management of natural resources.

2.2. Review of the literature

2.2.1. The PAs

2.2.1.1. The functions of protected areas

As tools for maintaining *in-situ* ecosystems, natural and semi-natural habitats, and viable populations of species in their natural environments, protected areas have multiple functions (from Parks for Life) (from PATRICK, 2009):

protection of species highly sensitive to human activities and disturbances;

maintenance of wild genetic resources important for medicine or for the reproduction of animal or plant species;

scientific research on spaces that resemble as closely as possible the original natural ecosystems;

soil and water conservation in erodible areas,

regulation and purification of water, notably by protecting wetlands and forests;

protection against natural disasters such as floods or storms; maintenance of important natural vegetation on poor soils and in sensitive areas;

provision of habitats for feeding, breeding or resting of species;

fundamental role in public education and awareness, especially in schools; protection of specific natural and cultural elements;

income and job creation through tourism.

Not all of these objectives can be developed in all protected areas. The purpose of this book is to show how priorities should be determined so that the site can best contribute to the objectives set for a particular protected area.

The common objectives of protected areas are, according to IUCN (2008):

conserve the composition, structure, function and evolutionary potential of biodiversity;

Contribute to regional conservation strategies (core reserves, buffer zones, corridors, staging areas for migratory species, etc.)

preserve the biodiversity of the landscape or habitat, species and associated ecosystems be large enough to ensure the integrity and long-term maintenance of the specified conservation targets, or be expandable to do so;

preserve forever the values for which they were created;

operate with the assistance of a management plan and a monitoring and evaluation program that encourages adaptive management;

have a clear and fair governance system;

preserve significant landscape features, geomorphology and geology;

Provide regulatory ecosystem services, including buffering against climate change impacts;

To conserve natural and scenic areas of national and international significance for cultural, spiritual and scientific purposes;

distribute benefits to local and resident communities in accordance with other management objectives;

provide recreational benefits consistent with other management objectives;

facilitate low-impact scientific research activities and ecological monitoring that is linked and consistent with the values of the protected area;

Use adaptive management strategies to gradually improve management effectiveness and governance quality;

help provide educational opportunities (including management approaches);

help gain general support for protection.

2.2.1.2. Rationale for establishing protected areas

Protected areas have been established to (From PATRICK, 2009):

1° species

It is recommended that protected areas be established to meet the following requirements for species preservation:

threatened species on the IUCN red list, with a special focus on critically endangered and endangered species;

endemic species, with a high priority for critically endangered and endangered species, found mainly at one site;

significant assemblages of gregarious species;

species important for conservation development and management (e.g., indicator species);

wild species ancestors of domestic or cultivated species.

2⁰ habitats and ecosystems

It is also recommended that protected areas be established to meet the following conditions for the preservation of :

viable representations of terrestrial, freshwater or marine ecosystems;

irreplaceable habitats and ecosystems (habitats or ecosystems that have unique characteristics such that no other area can be conserved in their place and that still retain these characteristics;

large, intact or unfragmented natural areas;

natural habitats with a high level of threat;

habitats necessary for the survival of viable populations of migratory species;

sites with a biodiversity useful for humanity;

sites providing services, such as hydrological functions, coastline and soil protection, breeding opportunities for economically valuable species;

Sites with species that are valuable economically or because of their genetic heritage (e.g., food, wood supply, medical and scientific research);

sites and species of particular socio-economic value (sacred sites, charismatic species, sites offering recreational and contemplative opportunities, landscapes of great beauty).

2.2.1.3. Protected areas regulations

Protected areas are governed by the laws of each State, with names that may vary according to national legislation. It is up to the States to create national networks of protected areas and to provide them with the means to implement a genuine conservation policy. Depending on the case, management may be decentralized or entrusted to non-state structures, such as NGOs, local communities or the private sector.

To ensure that everyone is talking about the same thing, IUCN has defined six broad types of protected areas based on the above objectives (IUCN WCPA protected area classification, updated from DUDLEY, 2008):

integral protection (e.g. a: integral nature reserve, b: wilderness area);

ecosystem conservation and recreation (e.g., national park);

conservation of natural features (e.g. natural monument);

conservation through active management (e.g., habitat/species management area);

Landscape/seascape conservation and recreation (e.g., protected landscape/seascape);

sustainable use of natural ecosystems (e.g., managed natural resource protected area)

These categories provide a common and universal language for understanding the world's protected areas beyond their different names. They provide a framework for collecting, manipulating and disseminating data on protected areas at the international level, and thus facilitate analysis, global and regional assessments and comparisons between countries.

The typology of these categories makes explicit the differences between the main types of protected areas and in management approaches. They recognize different management arrangements and different types of governance.

Despite their numbering, these categories do not imply a simple hierarchy in terms of quality, significance or preservation.

2.2.2. Armed conflicts

2.2.2.1. Origin of conflicts

Conflict occurs when two people or a group of people pursue divergent or seemingly divergent interests. Conflict is natural and inherent in human life. Conflict arises from natural disagreement resulting from people or groups of people differing in attitudes, beliefs, values or needs. It can also arise from past rivalries and personality differences. Conflict occurs when 2 or more parties pursue interests that are or appear to be incompatible. There are five main sources of conflict between two parties. Knowledge of these causes can help establish what is expected of one or both parties to resolve the situation (CHARLINE and CHARLES, Sd).

Table 1: Main causes of conflicts

ROOT OF THE CONFLICT	POSSIBLE CAUSES
Values	Differences in lifestyle, ideology and religion; Have very different value scales; Seeing a difference between a person's behavior and what they say their values are.
Resources	Two or more entities competing for perceived limited resources; Perceived unequal control, distribution or ownership of resources; The presence of geographical, physical or environmental factors that impede cooperation.
Interpersonal factors	Poor general knowledge of others; Stereotypes; Untested assumptions about the other; Disputes that have never been resolved; Previous disastrous encounters with the other party.
Interests	Competing needs, wishes or desires; Fundamental, procedural or psychological interests that are perceived as competing
Facts	Lack of information; Misinformation.

Source: CHARLINE and CHARLES, (Sd).

The above table shows the main causes of conflict in our Park. The ideal would be to find a way out of this crisis. This would be the subject of another study.

Table 2: Different types of conflicts and their causes.

TYPES OF CONFLICTS	CAUSES
Park/population conflicts	Felling of boundary trees; Depredation of crops by the Park's animals; Relocation / Evacuation of indigenous populations without compensation; Insufficient sharing of benefits between the PNKB and the population; Poverty of the population; Ecological corridor of the Park crossing the chiefdom of NINDJA.
Conflicts over access to natural resources	Extension of the park without consulting the population or accompanying measures; Mining of minerals in the park; Installation of fields in the park Wood cutting (construction, heating, embers); Poaching; Fishing with creels and use of toxic vegetation products.
Intra and inter-institutional conflicts	Arming of the herdsmen in the Park; Presence of armed bands in the Park (FDLR); Use of the Park for political purposes; Distribution of property titles in the Park other state services; Corruption; Weakness of the legal framework that does not favor co-management of the Park with the population; Non-publication of the law; Absence of national community conservation policy; Illegal traffic in the Park; Non-compliance with the legal texts governing the ICCN; Malfunction of the ICCN; Divergence and conflict between certain public services (ICCN, Rural Development, Mines, Environment, Land Titles, Agriculture and Livestock Division); Poaching by PNKB agents; Deforestation by some PNKB agents; Exploitation of minerals by some PNKB agents; Absence of the materialization of the limits.

Source: General Management Plan of the PNKB 2009-2019

This table traces the different types of conflicts and their causes in the protected areas of Kahuzi Bienga National Park. In the lowland area, illegal exploitation of columbite-tantalite (coltan), gold, and cassiterite has resulted in an influx of thousands of people, with a consequent increase in poaching and habitat destruction. In the Nzovu and Itebero sectors, some elements of the FARDC are also involved in the plundering of the natural resources of the PNKB. They destroy wildlife and are complicit in mineral exploitation (ICCNPNKB, 2007).

After a long absence of the ICCN in the large part of the park's base elevation, initially less protected, with the result that illegal subsistence activities such as mineral exploitation have increased, so the local population has continuously opposed the control of the park authority. This is an issue that nations should address, given that the park has become *a World Heritage Site.*

2.2.2.2. Armed conflicts in the DRC

A. Reasons and stakes of the armed conflict in DRC

The history of armed conflicts in the DRC has caused a great deal of discussion in the search for the real reasons and stakes of this monotony of wars, for the Rwandophone populations, otherwise known as Banyamulenges, these are wars for the defense of their ethnic identity, because it is necessary to defend oneself when one is part of a minority that is discriminated against and exposed to extermination. But this thesis does not convince anyone, because out of 450 Congolese tribes, no tribe is either a minority or a majority.

This suggests that the regime in Kigali is manipulating the Congolese Tutsis by throwing them into a merciless war in which all the spoils go to Rwanda, which in turn reports to the multinationals. It is certain that the different wars have caused the death of tens of thousands of Banyamulenges used as racehorses in these conflicts. This is the reason for the revolt of Tutsi General Masunzu. For Rwanda, his version is that it must secure its borders against attacks by the FDLR and ex-genocidaires interahamwes.

This hypothesis is also no longer appropriate because Rwanda, under the label of the RCD, controlled the East of the country for five years, but it was more concerned with the plundering of wealth than with tracking down Hutu fighters. Congolese civil society, on the other hand, is unanimous that it is the covetousness of wealth by multinational firms that use the neighboring countries Rwanda and Uganda. In addition, the permanent and murderous insecurity justifies the process of balkanization of the country. This second version causes tension among the local population.

The thesis of the civil society seems to be close to the realities of several publications and testimonies of the above-mentioned official figures of the Western States. Finally, it should be said that the reasons for these wars are holistic.

B. The main actors of the armed conflicts in DRC

Among these actors we find:

1° External actors are external forces that interfere in a conflict fueled by aggravating factors that include ethnic dialectics, conflicting economic interests and a demographic situation characterized by high densities (KIPPENBERG, et al, 2009; HUMAN RIGHTS, 2009).

2° The internal actors are the Congrès National pour la Défense du Peuple (CNDP), the M23 rebels, the Front Démocratique de Libération du Rwanda (FDLR), the Mayi Mayi resistance patriots, the Lord's Liberation Army (LRA), the Forces Armées de la République Démocratique du Congo (FARDC) the United Nations Mission for the Stabilization of the Congo (MONUSCO), neighboring countries (Rwanda, Uganda and Burundi), multinational firms, the ADF NALU, the Patriotic Forces for the Liberation of the Congo (FPLC), the "Raia Mutomboki in Swahili which means the population is revolting."[1]

2.2.3. Armed conflicts and their effects on PAs

2.2.3.1. General analysis of the situation

Habitat destruction and loss of wildlife are among the most widespread and severe effects on PAs caused by armed conflict, which may be for strategic, commercial, or subsistence reasons. For example, vegetation may be cut, burned, or defoliated to increase troop mobility and visibility. In Rwanda in 1991, the army mowed a 50-100 m wide strip through the bamboo forest adjacent to the Virunga volcanoes in order to reduce the risk of ambush along a major trail (KALPERS, 2001).

In areas where fighting is taking place, troops regularly hunt large numbers of large mammals for food. This practice can have disastrous consequences for wildlife populations, especially if military activities continue over a long period of time.

---

[1]   http://www.unmultimedia.Org/radio/english/2012/06/un-human-right:s-chief-fears-more   rapes-killings-in-congo-by-m23/ (accessed 03/23/2017; POURTIER, 2009; URL: http://echogeo.revues.org/10793: HumanRi ghts Watch(HRW) ;
http://wilkipedia/wiki/,armée of resistance of the lord).

When displaced people are temporarily resettled, they cut vegetation for agricultural purposes or to obtain firewood. Such practices quickly lead to deforestation and erosion. As refugees and displaced persons are often resettled in marginal and vulnerable ecological areas, environmental recovery is limited. Political instability during wartime often has the immediate consequence of making it impossible for residents to cultivate crops. To survive, they are gradually forced to turn to wild foods such as meat and food plants. At the same time, displaced people collect firewood, food plants and other local natural resources.

This is why it is important to establish specific management of protected areas during armed conflicts. If nothing is done in this sense, biological diversity can be seriously threatened. Therefore, as soon as a conflict breaks out and threatens the natural resources of protected areas, a management adapted to the situation must be adopted. A decision to cease conservation activities may have to be taken if the situation becomes too unstable, and conservation staff may be forced to leave their duties and flee, with some at risk of being killed.

Senior staff are often the first to leave, and senior staff with access to project funds or vehicles are a prime target. On the other hand, senior staff may belong to an ethnic or religious group targeted by political enemies (PLUMPTRE *et al.*, 2001).

Thus, the relevant authorities (or conservation agencies) responsible for managing protected areas remain in charge of managing them in times of conflict. In the absence of a conservation sector, the relief and development sectors will be able to mitigate adverse environmental effects by integrating environmental considerations into their programs.

Maintaining a presence in times of conflict to ensure support for conservation and natural resource management by all means is one of the greatest challenges facing conservation organizations. Some operational aspects stand out as priorities (PATRICK, 2009):

> close collaboration with local civil, military and traditional authorities at all levels;
>
> the development of diplomatic negotiation techniques;
>
> a certain flexibility with regard to power movements.

Conservation organizations, those working in a more diverse environment, may already have adopted a multi-sectoral approach. Their adaptation may be easier than that of organizations working exclusively in protected areas. Considerable changes may occur during armed conflict in the sources and extent of funding for activities. Measures to mitigate the adverse environmental effects of armed conflict can be implemented in the following three phases:

prevention and preparedness: taking a proactive approach before a crisis occurs can significantly contribute to achieving conservation goals during and after conflict;

Adaptation and mitigation: during the crisis, adaptation strategies help to mitigate the negative effects of the conflict on the environment and to make full use of the opportunities that may arise, this phase includes the duration of the conflict and the transition to peace;

Post-crisis: once the conflict is over, longer-term recovery, rehabilitation and reconstruction programs must be developed and implemented to promote appropriate action for environmental protection.

Two factors are essential to maintaining a presence in times of conflict: committed personnel and adequate funding. If it becomes necessary to leave a site, efforts should be made to remain active in the area. Even when a conservation organization is forced to leave a particular site, it must assess the level of security and develop security measures. The threat assessment requires five types of data:

the nature of the threat;

the situations in which this threat may materialize;

the level of the threat;

the potential evolution of the threat;

the causes of the threat, e.g., criminal act or banditry, direct threats to staff or the conservation organization, or indirect threats that unintentionally affect staff or the organization while a third party is targeted.

On site, during the armed conflict

The tasks to be accomplished are as follows (PATRICK, 2009):

provide financial, logistical and moral support to staff;

help staff stay healthy;

Hire competent, specialized and preferably local personnel;

Install appropriate equipment for emergency communications;

develop communication plans and procedures;

Assess training needs and undertake the required training;

Ensure that staff understand and believe in their mission;

identify who will benefit from training and how that training is funded;

promote joint planning with relief and development organizations;

integrate humanitarian issues into programming;

be neutral;

respect its mission;

open the communication channels;

develop partner capacity;

strengthen collaboration between headquarters and field staff;

Determine specific roles and responsibilities and identify lead agencies in each area;

Involve local communities in conservation interventions related to refugee absorption;

monitor the effectiveness of interventions;

be flexible;

Consider the relationship between the military and the local community;

Work with appropriate authorities, organizations and communities to mitigate the impact of refugees crossing the border;

to seek opportunities for cross-border collaboration during the post-war rehabilitation period;

Seek opportunities to promote peace at the local level through transboundary conservation initiatives and vice versa;

consider the adoption of a code for transboundary protected areas in times of armed conflict;

develop partnerships with the various responsible authorities while maintaining neutrality;

analyze the environmental impacts of humanitarian programs;

Ensure donor coordination; provide emergency funds.

These tasks require skills, abilities and resources: leadership skills;

conflict resolution and negotiation skills;

ability to communicate;

telecommunication equipment, including satellite radio/mobile phone; sanitary equipment, medicines for first aid.

2.2.3.2. Presentation of the impacts of armed conflicts on the PAs of the DRC

For plant reserves: These are the ecosystems (ANONYMOUS, Sd)

During wars, there are movements of displaced populations that often lead to deforestation, which seriously hinders the economic and social development of the DRC. They contribute to the degradation of production systems, deterioration of the environment, loss of biodiversity, increase of greenhouse gases, decrease of agricultural yields and exacerbation of poverty.

Thus, the influx of displaced persons commonly called "refugees" is accompanied by a minimum of vital activities: installation and occupation of forests and/or savannahs, search for wood for the construction of their dwellings, but all these activities weigh down the biosphere, especially the species.

It is necessary to popularize the culture of peace, to highlight the ethical responsibility of man on his environment. The culture of peace and ethics emphasizing the notion of ecological constraints inflicted by man on his environment as a provider of resources and the promotion of life through ecological culture, education and mesological ethics. In short, we need a place for the environment that can ensure the responsible and convivial cohabitation between man and nature.

For animal reserves: they are national parks (ANONYMOUS, Sd)

Animal species are suffering. The war in the east, north and south has eliminated rare animal species such as the Okapi, the gorillas, the dwarf chimpanzee or bonobo, and the northern white rhinoceros. The parks were occupied by people displaced by wars. The aggressors transferred the rare species to their respective countries (Rwanda, Uganda, Burundi). The United Nations Commission of Inquiry has made damning revelations about the plundering of the DRC's wealth by both the aggressor countries and the rebels.

Of the eight national parks in the DRC, some of which are classified as UNESCO World Heritage sites, five are located in the east of the country. These have been progressively included in the list of endangered heritage.

Virunga National Park 1994, in North Kivu;

Garamba National Park, Orientale Province, in 1996;

Kahuzi-Biega National Park in South Kivu, 1997;

Okapi Wildlife Reserve in 1997 in the Maïko National Park;

Salonga National Park in 1999, Ecuador.

Unfortunately, the ecological damage resulting from armed conflict and illegal resource exploitation is considerable in this unique environment. "The combination of exploitation activities and ongoing conflict has effectively eliminated administrative control over the parks and led to the militarization of many of them, including Virunga, Kahuzi-Biega, and Okapi reserves. These parks occupy a strategic position along the eastern border of the DRC, and are regularly used as a crossing point by armed forces to enter eastern Congo.

They are also the site of violent fighting between local rebel forces and armed groups that occupy parts of the parks almost permanently. They regularly poach elephants for the ivory trade, game and rare species, and plunder forest resources. "The migratory patterns of many wildlife species have been significantly disrupted, creating lasting problems in repopulating some park areas or maintaining population balance in others.

The local populations also settle or resettle in the parks to ensure their subsistence through fishing, poaching and intensive tree felling. Due to multiple circumstances and difficulties they face, different compulsory cultures are developed without distinction of gender in age, the output in the mines is pushed to the maximum in the Congolese national parks and yet inscribed on the world heritage list.

Forced labor has increased, especially of women and children, as the conscription of men into the army has created a labor shortage. Biosphere reserves have been destroyed in a long war, particularly despite the fact that UNESCO promotes the identification, protection and preservation of cultural and natural heritage considered to be of outstanding value to humanity.

It should be noted that privatized warfare is the work of investors who plunder, at the cost of arms protection alone, resources that would otherwise be more expensive. The operation often takes place under the diplomatic umbrella of the country that is in charge of "boiling down the so-called international law". The deterioration of the national environment continues in full view of everyone, even the leaders remain silent on this issue. The national parks are so ravaged despite their inscription on the list of the world heritage.

CHAPTER THREE
SURVEY, ANALYSIS AND INTERPRETATION OF RESULTS

3.1. PRESENTATION OF THE STUDY ENVIRONMENT

The present study is conducted in the KBNP in the Lulingu sector in Shabunda territory as an ecological entity as well as in the surrounding localities, South Kivu Province, Democratic Republic of Congo.

3.1.1. Geographical location of the PNKB and context of its creation

The PNKB is located in the East of the DR Congo in the province of South Kivu, it extends from the basin of the Congo River near Itebero-Utu to its western border in the North-East of Bukavu with an area of 600,000 hectares. Because of its geographical location (1°36'-2°37' South latitude and 27°33'- 28°46' East longitude) and the variable relief, a rich fauna and flora can be found here, the most spectacular representative of which is certainly the Eastern Lowland Gorilla.

This is one of the areas in Africa where the transition from lowland to montane rainforest has remained largely intact, according to (WHITE, 1983) *in* FISCHER (1993). The KBNP borders the territories of Kabare, Kalehe - Bunyakiri, Walungu, Shabunda in South Kivu and Walikale (in North Kivu).

Regarding the context of the creation of the PNKB, this park was created to safeguard the typical mountain forest of the Kahuzi and Biega mountains and to protect the eastern lowland gorilla subspecies, Grauer's gorilla (*Gorilla gorilla graueri*).

Besides this animal species, two others attract the curiosity of the managers of this Park.

These are the Forest Elephant (*Loxodonta cyclotis*) and the Chimpanzee (*Pan troglodytes schweinfurthii*). These three species constitute the flagship animal species of the PNKB.

The creation of the PNKB took place in 1937 by the ordinance law number 81/AGR of Mr. RYKMANS, governor general of the Belgian Congo on a surface of 75000 ha and under the denomination of the integral forest zoological reserve in the region of the mount Kahuzi incorporating also 3 blocks of breedings of the settlers.

On November 30, 1970, the status of the reserve was lifted by Presidential Order No. 70/316 and the reserve was transformed into the Kahuzi-Biega National Park with a reduced surface area of 60,000 ha following the non-integration of certain settlers' livestock blocks into the Park.

In July 1975, on the proposal of Mr. Adrien DESCHRYVER, a curator considered to be the founder of the KBNP, the area of the latter was increased from 60,000 ha to 600,000 ha by presidential order n°75/238, thus incorporating again certain farms of the colonists and the integration of a major part of the forest in the low altitude. Following this extension, the PNKB is made up of two regions with different characteristics, the lowland block and the highland block linked by the Nindja connectivity corridor (SOURNIA et al., 1998).

The importance of the PNKB was recognized in 1980 by UNESCO, which granted it the status of a World Heritage Site.

3.1.2. Geomorphology, Geology and Soil

According to RUNG (1992), most of the KBNP is located in the central Congo Basin. This area consists of steep mountainous terrain cut by deep valleys. According to the same author, Mount Kahuzi and Mount Biega are not, as is often assumed, volcanoes, but are composed of Precambrian metamorphic rock formations.

Because of its geographical location, the topography is very varied. From the Central Congo Basin, the relief rises continuously to the western ridge of the Central African Graben of the Western Rift. In the low-lying part of the park, the lowest point is about 700 m above sea level, the highest point is near Bunyakiri (Mount Kamami) at about 1700 m above sea level. In the high altitude part, Mount Kahuzi (3,308m) dominates the mountain range.

At the bottom of the Graben is Lake Kivu at about 1400 m elevation. Heavily folded Precambrian metamorphic Burundian rocks and stones from the Graben ridge predominate the eastern region. However, sandy sediments and earthy stones are found in the west. The region around Lake Kivu is really influenced by Tertiary tectonics, linked to Quaternary volcanism. According to BUSANE (2006), the soil in PNKB is volcanic, clayey and sandy.

3.1.3. Climate and hydrography

The KBNP has a sub-equatorial and mountain climate. In the lower parts (700 to 1700 m above sea level), the climate is uniformly warm during the day and throughout the year. The average annual temperature is 20.5°C in Irangi. Rainfall is very high, but not evenly distributed throughout the year. There are two rainy seasons separated by short dry seasons: one from May to June, the other from October to December.

On the other hand, the mountainous region (dominated by Mount Kahuzi: 3308m above sea level and Mount Biega: 2790m above sea level) is dominated by an Afro-alpine climate with night frost on the peaks. During the day, there is abundant cloud cover and heavy rainfall, especially in the afternoon and evening (BUSANE, 2006).

HEDBERG (1957) summarizes this situation by *"summer every day winter every night"*. There is a close relationship between altitude and average temperature. The average annual precipitation amounts to a maximum of 1900 mm with a dry season from June to August. The average monthly temperature is around 15°C, according to GRIFFITHS (1972).

The KBNP is crossed by a large number of rivers. The main ones are the *Luka, Talya, Utu, Duma, Kamatunga, Tshomoka and Kanzuzu* rivers in the north; the *Mabu, Numbi, Katwalo, Mahungu* and *Lugulu* rivers in the center; and the *Nyakagera* and *Lubimbe* rivers in the south (ICCN, 2000).

3.1.4. Flora and vegetation

The KBNP is subdivided into two zones linked by a narrow corridor: the montane rainforest (or Afro-montane forest) on the one hand and the planitic rainforest (Guineo-Congolese, relatively humid type) on the other.

MUHIGWA et al (2007) report that the KBNP covers continuous tropical vegetation, ranging in elevation from 600 to 3,308 m. The vegetation continuity of KBNP is unique, not only for Congo, but for all of sub-Saharan Africa. This situation ensures genetic exchange between low and high altitude populations of large mammals.

Given the small area of highland forests and their isolation in the overpopulated and fully cultivated region, this exchange is the only guarantee for the survival of these threatened large mammal populations. MÜHLENBERG et al (1994) identify 6 major vegetation formations in the KBNP:

mountain rainforests and secondary vegetation;

High mountain rainforests and secondary vegetation;

swamp forests;

bamboo forests and secondary vegetation;

subalpine heaths;

marshes and peat bogs.

SOURNIA et al. (1998) also note that the plant formations of the PNKB are distinguished into several types of plant environments:

the ombrophilous forests of Mountain extending from 900 to 2300 m of altitude and the dominant species are *Albiziagummifera, Parinariexcelsum* and *Chrysophyllum gorungosanum*. This forest is transformed into secondary forest after the passage of the man presenting itself as a shrubby environment constituted by the undergrowth and lianas, called forest to *Hagenia* and savannas included;

High mountain rainforests, also called *Podocarpus* forests, are found above 2300 m, characterized by the species *Podocarpus usambarensis, Syzygiumguineense* and *Psychotriamahoni*;

Swamp forests where species have developed stilt roots and pneumatophores, dominated by *Syzygiumrowlandii, Podocarpusambarensis, Agauriasalicifolia, Anthocleistagrandiflora, Cyperus* swamps are frequented in the rainy season by gorillas that feed on its pith and fresh leaves;

The bamboo forests are located between 2,300 and 2,600 m, with an annual rainfall of at least 2,000 mm. We note an extension of these forests which is the direct consequence of the anthropic clearings;

heather formations above 2600 m where trees can no longer grow.

3.1.5. Wildlife

Although the Park was created to protect the Eastern Lowland Gorilla, a species that exists only in eastern DR Congo between the Lubutu River in the north, Lubero River in the northeast, and Fizi River in the south, non-gorilla forest wildlife is also well represented.

A variety of other mammals such as elephant, buffalo, hylocher, bushpig, various cats, genets, mongooses, otters, bongo (and many other antelopes), chimpanzee, pangolins, various galagos and many other monkeys are found in the park, as well as a variety of birds, reptiles and amphibians living in the park and, at lower elevations, in the surrounding forests (MÜHLENBERG et al, 1994).

PNKB is therefore an important site for its forest fauna and avifauna. The forest fauna is represented by a variety of mammals, birds, reptiles, amphibians, etc. The same fauna exists in the western extension of the park where the presence of high altitude avifauna is also reported.

Here is the general situation of the fauna of the PNKB

This situation is as follows according to ICCN-PNKB (2009):

Amphibians: Twenty-nine (29) species of amphibians have been recorded within the boundaries of the national park;

Reptiles: the inventory of this group is far from being exhaustive but 44 species are known;

Birds: 10 bird species of great importance are present in KBNP, of which 4 have an altitudinal range of lowland forests. Overall, the number of known bird species in the KBNP is 260;

Mammals: In the PNKB there are about 134 species of mammals.

3.1.6. The Lulingu sector

Lulingu is a locality of the Bamuguba-Nord grouping, Bakisi chiefdom in the **Shabunda territory** in the province of South Kivu, located 40 km from Katchungu and 100 km from Shabunda:

South latitude: 02 20' 32.7'';

Longitude East: 27 36 37.5''.

With the nearby villages (within a 10 km radius), it has about 35,000 inhabitants, or 7,000 households. In addition, there are many displaced persons who come to Lulingu to seek refuge. The number of displaced persons varies greatly depending on the intensity of the unrest in the region. The last assessment, made in early 2012, estimated that approximately 4,000 households (or +/- 20,000 people) had temporarily settled in Lulingu. This influx of refugees obviously brings its share of difficulties of all kinds in the areas of health, nutrition, education, etc.. All problems that the authorities and inhabitants of Lulingu are unable to manage, as they themselves are already living in more than precarious conditions.

The area of Lulingu is covered with primary forests, located in the north, and serves as a refuge for all kinds of rebellions. It is from this area that most of the refugees who come to take shelter in Lulingu come. This forest area is also included in the 600,000 hectare Kahuzi-Biega National Park (two ancient extinct volcanoes) where many endangered animal species remain, including lowland gorillas. Thus, the Lulingu forest area is part of the KBNP.

Figure 1 below shows the map of the PNKB and sectors in its old configuration.

IUCN, 2010.

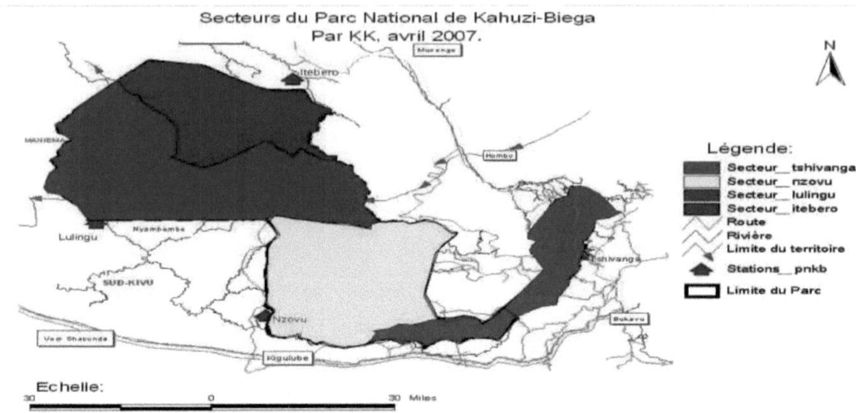

Figure 2 below shows the map of the PNKB and its sectors in its new configuration.

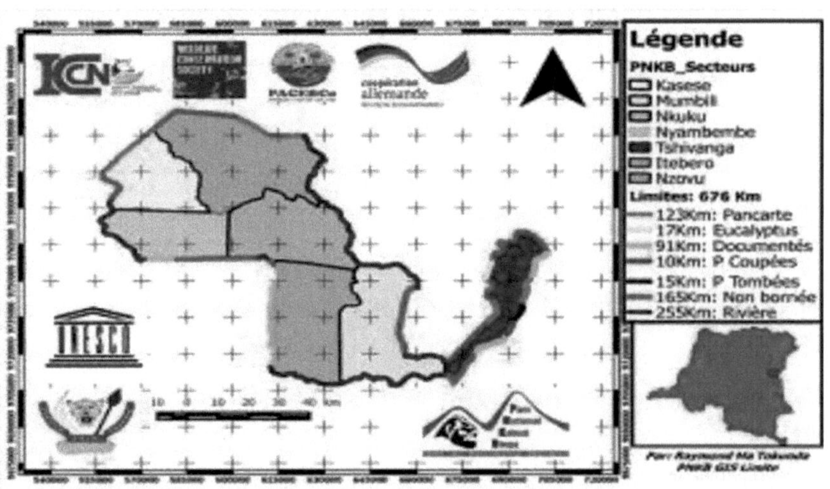

Source: ICCN, 2016

Map of the PNKB and its sectors in its new configuration, received from the archives of the ICCN Office 2016. It is in vogue for its days.

3.1.7. Populations living in the vicinity of the PNKB

Around the Kahuzi-Biega National Park live human populations of different communities. These are the riparian populations of the KBNP. These communities are constituted on the one hand by different ethnic groups of Bantus: Shi in the territories of Kabare and Walungu, Havu and Tembo in the territory of Kalehe, Tembo and Rega in the territory of Shabunda, Rega and Nyanga in the territory of Walikale in its North-Western part and on the other hand, by pygmies called indigenous populations.

Thus, human activities are intense both within and at the edge of the Park due to the existence of these two population groups in the vicinity (ICCN, 2007). In the past, both groups were always nomadic. While communities in the first group preferred a sedentary lifestyle based on income from agriculture and livestock, those in the second group preferred a nomadic lifestyle based on hunting and gathering. While the former changed their villages after a few years and according to the exhaustion of agricultural land around the village, the latter changed their camp as soon as the results of hunting and gathering weakened. This is what the ecologist calls the temporary "over-hunting" in a specific area and after a few weeks. It is believed that these two groups have always lived in harmony, exchanging their products for the benefit of both (VANSINA, 1990).

It seems clear that these changes have contributed to a profound change in the relationship between farmers and "forest peoples. While farmers became increasingly involved in rural changes, hunter/gatherers, forest dwellers or "pygmies" were increasingly marginalized.

According to NGUIFFO (2003), pygmies are the best known and most vulnerable of the forest populations. Their way of life is closely linked to the forest, from which they draw food (game, fruits, bark, roots, etc.) and the products of the traditional pharmacopoeia of which they are reputed to be great connoisseurs. The forest constitutes their natural living environment, in which they continue, for the most part, to practice nomadism.

Some have tried to adopt the farm life style without being able to achieve the same productivity as the farmers and without being able to play a role in the social debate and/or decision-making processes.

According to ICCN-PNKB (2009), PNKB is located in one of the most densely populated regions of the DRC, with a current density of approximately 400 inhabitants per km in the high altitude part^2 and a conventional density of 7.5 inhabitants per km^2 in 1980 in the low altitude part. The majority of these inhabitants are poor, with an income below the conventional poverty line of 1 dollar per day per person.

This human concentration has consequences on the fauna and flora as well as on the habitats of the park through the extraction of its natural resources (mining and agricultural exploitation, illegal exploitation of wood-energy, etc.) and the degradation of the fauna through poaching (HART et al., 2007).

3.2. DATA COLLECTION METHODOLOGY

3.2.1. Drawing of the sample

At present, the data in our possession show that in the Lulingu region, there are about 55,000 inhabitants or 11,000 households.

To determine the sample size of our study, we used the total population of our study area which is 55,000 inhabitants by applying the technique of YAMAN (1970) from the course of scientific research given by Professor

ILUNGA LUTUMBA with reference to the following formula:

$n = \frac{N}{1+N(e^2)}$ with a degree of precision or margin of error of 9% or (0.09).

Where: n=sample size; N=total number of the population and e =degree of precision or estimated margin of error at 9%.

Thus, the calculation is as follows to determine our sample size considering that the total population of the study area is 55,000 inhabitants.

$$n = \frac{55\,000}{1+55\,000 \times (0.09 \times 0.09)} = \frac{55\,000}{1+55\,000 \times 0.0081} = 123,180$$

n = 123,180 ~ 123 people to which we added 7 additional people for reasons of ease of calculation, which makes the sample of our study 130 people to investigate in our study environment.

Considering the WHO formula which is as follows:

$$e = \frac{a \times 100}{b} = \frac{130 \times 100}{55\,000} = \approx 0,236\,\% \approx 0,24\%$$

Where e = sample, a = number of respondents, b =

Total population, our sample represents 0.24% of the total population in our study setting.

3.2.2. Techniques and methods of data collection

3.2.2.1.Materials

3.2.2.1.1. Biological materials

The biological material used in this research consists of the population of the locality of Lulingu in the Bamuguba-Nord grouping, Bakisi chiefdom in the **Shabunda territory** in the province of South Kivu, to whom we submitted our survey questionnaire.

3.2.2.1.2. Technical equipment

For the success of this work we used as technical materials: the logbook and the pen which allowed us to record the various data during our surveys, the backpack for the transport of the survey questionnaire and the survey questionnaire.

3.2.2.2.Research methods

According to MADELEINE and PINTO (1976), the method is constituted by the whole of the intellectual operations by which a discipline seeks to reach the truths which it pursues, to demonstrate it and to check it.

In this study, we used the following methods:

1. Comparative method

It allowed us to compare the descriptive features of the ecological status of the past with the current status of the PNKB. It was also useful in identifying the causes and consequences of the current degradation of the ecological state of the KBNP and how to act and by whom.

2. Descriptive method

It allowed us to accurately present all aspects of the KBNP. It also allowed us to describe the current ecological state of the PNKB through field observations. It allowed us to give the qualifying features of the ecological state of this park.

3. Historical method

It has given us a lot of knowledge about the past of the PNKB, therefore about the evolution of the general ecological state of the park, but also about the evolution of the influence of the socio-economic activities carried out by the riparian populations on the park over time.

4. Explanatory method

This was useful because it allowed us to explain our questions in order to have not only sufficient data but also well structured data in relation to our research topic.

5. Synthetic method

This method helped us because we had to synthesize the data obtained in the field in order to present them well and make them easy to understand.

3.2.2.3.Research techniques

Techniques are tools at the service of the method. No field research can be done without the use of techniques. In this study we used the following techniques:

1. Survey technique

The survey was made possible by a survey questionnaire that was administered to 130 people in all sub-locations of Lulingu. The survey was conducted in the morning or evening when the respondent was at home.

The questions of the survey questionnaire were submitted to the respondent in Kilega and sometimes in Swahili. This survey took place over a period of 30 days from 01 to 05/ to 30/6/2017. The survey questionnaire to be submitted to the respondents, a copy of which is attached, consisted of closed questions. This type of questionnaire consists of a set of questions asked but with several answers provided and from which the true answer is chosen. These questions are called "closed-ended" because they do not allow the respondent to add unnecessary comments to the question asked, which facilitates the coding and analysis of the results.

2. Observation technique

Observation is one of the important techniques of scientific research because nothing can replace direct contact of the investigator with his field of study and no technique is capable of suggesting so many new ideas. It is difficult to imagine a serious attitude of behavior in which observation plays no role.

Thus, to supplement the information we collected in the field during the surveys, we undertook observations of the daily activities of the inhabitants of the locality of Lulingu and more particularly those of the FRDC and the other armed groups, activities carried out in and around the PNKB.

3. Documentation technique

It is a technique that helped us to search and interrogate the documents in order to obtain reliable information on our subject. To carry out this work, we used books, TFC, Memoirs, Syllabus and the Internet.

4. Data processing

It is an approach that consists of a means of gathering and analyzing the data collected in the field in order to make them numerical or quantified. To achieve this, the data collected in the field was processed and organized on the computer using Excel software to make it quantitative. The interpretation of these data allowed us to verify the hypotheses and achieve the objectives of the research.

3.2.3. Summary of results

In order to verify our hypotheses and to reach our objectives, but also to be aware of the reality in the field, we have developed a survey questionnaire that could provide us with useful information in relation to our research theme.

The survey questionnaire was intended for households in the Lulingu area and was designed to collect data on armed conflicts in the area, their causes and effects on the KBNP, in order to propose solutions that could help put an end to these conflicts and thus protect the KBNP and its resources.

3.3. Analysis, Interpretation and Discussion of Results

3.3.1. Characteristics of the respondents

1. Gender of respondents

The results concerning the gender of the respondents are presented in the table below:

Table 3: Distribution of respondents by gender

Type	Workforce	%
Male	104	80
Woman	26	20
Total	**130**	**100**

Source: our investigations in the field

This table shows the distribution of respondents by gender. Overall, 80% of men were surveyed in our study area, compared to 26.6% of women. This is justified by the fact that men were much more available to answer our questions and were much more likely to have sufficient knowledge about our research topic.

2. Age range of respondents

The results concerning the age groups of the respondents are presented in the following table.

Table 4: Distribution of respondents by age group

Age in Range (years)	Workforce	%
18-30	49	37,6
31-50	64	49,2
50 and more	17	13
Total	**130**	**100**

Source: our investigations in the field

This table shows the distribution of respondents by age group. Overall, 37.6% of respondents are between 18 and 30 years old, 49.2% are between 31 and 50 years old, and 13% are 51 years old or older. The majority of our respondents are between 31 and 50 years old, with an average age of 41 years. This is justified by the fact that in this age group many respondents were available to answer our questions.

3. Marital status of respondents

The results concerning the civil status of the respondents are presented in the following table:

Table 5: Distribution of respondents by marital status

Civil status	Workforce	%
Married	98	75,3
Single	20	15,3
Divorced	7	5,3
Widow / Widower	5	3,8
Total	**130**	**100**

This table presents the distribution of respondents by marital status. Overall, 75.3% of the respondents were married, 15.3% were single, 5.3% were divorced and 3.8% were widowed. In view of these results, it is clear that married people were much more willing to answer our questions.

4. Household size in slices

The results for household size of the bracketed respondents are presented here:

Table 6: Distribution of Respondents by Household Size by Band

Household (in increments)	Workforce	%
1-5	43	33
6-10	64	49,2
More than 10	23	17,6
Total	**130**	**100**

Source: our surveys.

This table presents the distribution of respondents according to household size. It emerges that 49.2% of respondents have households with between 6 and 10 family members, 33% of respondents have households with between 1 and 5 family members and finally 17.6% of respondents have households with more than 10 family members.

5. Main occupation

The results concerning the main occupation of the respondents are presented.

Table 7. Distribution of respondents by main occupation

Occupancy	Workforce	%
Park Agent	20	15,3
State Agent	40	30,7
Merchant	28	21,5
Farmer	21	16,1
Hunter	4	3
Mason or carpenter	3	2,3
Humanitarian	3	2,3
Without	11	8,4
Total	**130**	**100**

Source: our investigations

This table shows the respondents according to their main occupation. In our sample, 30.7% of respondents were government employees, 21.5% were shopkeepers, 15.3% were park employees, 8.4% were humanitarians, and the other categories of respondents according to their main occupation came next. These results show that our sample is representative in terms of the occupation of the respondents.

3.3.2. Actual survey results

1. Respondents' opinions on the presence of one or more armed conflict actors in and around the KBNP in the Lulingu area

The results regarding respondents' opinions on the presence of one or more armed conflict actors in and around the KBNP in the Lulingu sector are presented in the following table.

Table 8. Distribution of respondents according to opinions on the presence of one or more conflict actors in and around the park in the Lulingu sector.

Presence of armed groups	Frequency	%
Positive opinion	130	100
Negative opinion	0	0
Total	130	100

Source: our research

This table shows the opinions of respondents on the presence of armed actors in and around the KBNP in the Lulingu sector. All respondents recognized 100% of the presence of armed conflict actors or armed groups in and around the PNKB in the Lulingu sector.

2. Categories of armed conflict actors currently present in and around the KBNP in the Lulingu area

The results regarding the categories of armed conflict actors currently present in and around the KBNP in the Lulingu area are presented in the table below:

Table 9. Distribution of respondents according to categories of armed conflict actors currently present in and around the KBNP in the Lulingu area

Armed groups present in the park	Frequency	%
FARDC only	20	15,3
RM only	31	23,8
FARDC and RM at the same time	79	60,7
Total	**130**	**100**

This table shows the types of armed groups present in and around the PNKB in the Lulingu sector. Three types of armed groups were identified by respondents: the FARDC and the MR for 60.7% of respondents, the MR only for 23.8%, and the FARDC only for 15.3% of respondents. These results show that the FARDC and the MR are the armed groups currently present in the park in the Lulingu sector.

3. Reasons for the presence of armed conflict actors currently present in and around the KBNP in the Lulingu area

The results regarding the reasons for the presence of armed conflict actors currently present in and around the KBNP in the Lulingu area are presented below.

Table 10: Distribution of respondents according to the reasons for the presence of armed conflict actors currently present in and around the KBNP in the Lulingu area

Reasons for the presence of armed groups in the park	Frequency	%
War strategy	48	36,9
Park RNs Conveying	73	56,1
Protection of the park and its NR	9	6,9
Total	**130**	**100**

Source: our research

The table above presents the distribution of respondents according to the reasons for the presence of armed conflict actors in the KBNP in the Lulingu sector. According to these results, the respondents recognized three types of reasons why armed groups are currently present in the KBNP, namely the coveting of the park's NRs for 56.1% of the respondents, war strategies for 36.6% of the respondents, and finally the protection of the park and its NRs for 6.9% of the respondents. From these results, it is clear that the actors of armed conflict are currently present in the KBNP in the Lulingu sector for reasons of coveting the park's NRs.

4. Elements underlying the presence of armed conflict actors currently present in and around the KBNP in the Lulingu area

The results concerning the elements underlying the presence of armed conflict actors currently present in and around the KBNP in the Lulingu area are presented in the table below:

Table 11: Distribution of respondents according to the elements at the root of the presence of armed conflict actors currently present in and around the KBNP in the Lulingu sector

Basis for the presence of armed groups in the park	Frequency	%
Weakness of the park monitoring service	20	15,3
Weakness in the restoration of the authority of the State in the environment	79	60,7
Ignorance of the importance of the park by these armed groups	31	23,8
Total	**130**	**100**

Source: our research

The table above shows the types of elements that underlie the presence of armed groups in the KBNP in the Lulingu sector. The respondents recognized three types of factors that underlie the presence of armed groups in the KBNP in the Lulingu sector, in order of importance: weakness in the restoration of state authority in the area (60.7% of the respondents), ignorance of the importance of the park by these armed groups (23.8% of the respondents), and finally, the weakness of the park's surveillance service (15.3% of the respondents). Thus, for the majority of respondents, the presence of armed groups in and around the KBNP in the Lulingu sector is justified by the weakness in the restoration of state authority in the area.

5. Respondents' opinions on the exercise of certain human activities in the PNKB by armed groups in the Lulingu sector

The results regarding the opinions of respondents on the exercise of certain human activities in the PNKB by armed groups in the Lulingu sector are presented in the table below:

Table 12: Distribution of respondents' opinions on the exercise of certain human activities in the KBNP by armed groups in the

Opinion on the exercise of human activities by these armed groups in the park	Frequency	%
Positive opinion	130	100
Negative opinion	0	0
Total	**130**	**100**

Source: our research

This table shows that all respondents (100%) recognized that armed groups present in and around the PNKB carry out human activities in the park.

6. Types of human activities carried out in the KBNP by armed groups in the Lulingu area

The results regarding the types of human activities carried out in the KBNP by armed groups in the Lulingu area are presented in the table below:

Table 13: Distribution of respondents according to the types of human activities carried out in the KBNP by armed groups in the Lulingu sector

Human activities undertaken by armed groups in the park	Frequency	%
Poaching of animals	48	36,9
Illegal logging of timber and PNFLs	8	6,1
Illegal extraction of minerals	63	48,4
Agropastoral exploitation	13	10
Total	**130**	**100**

Source: our surveys

This table gives us the human activities undertaken by the various actors in the armed conflicts in the KBNP in the Lulingu sector. In total, four types of human activities are carried out by armed groups in and around the park in this sector, in particular, in order of importance, illegal mineral extraction (48.4% of respondents), animal poaching (36.9% of respondents), agro-pastoral exploitation (10% of respondents), and finally, the illegal exploitation of timber and LFN (6.1% of respondents). In view of these results, we note that illegal mineral extraction and animal poaching are the two most important human activities carried out by armed groups in the KBNP in the Lulingu sector.

7. Ecological effects of the presence of armed conflict on the park in the Lulingu area

The results regarding the ecological effects of armed conflict on the park in the Lulingu area are also presented in the following table:

Table 14: Distribution of respondents according to the ecological effects of armed conflict on the park in the Lulingu sector

Ecological effects of armed conflict on the park	Frequency	%
Loss of habitat for park wildlife	43	33
Pollution of the park ecosystem	20	15,3
Continued loss of wildlife species and other NR in the park	67	51,5
Total	**130**	**100**

Source: our research

The results of this table show the ecological effects of human activities on the KBNP carried out by the actors of the armed conflicts in this park. The respondents identified three types of ecological effects of the activities undertaken by these armed groups in KBNP, namely the continued loss of fauna and flora species and other NR in the park (51.5% of the respondents), the loss of habitat for the park's fauna and flora (33% of the respondents), and finally the pollution of the park's ecosystem (15.5% of the respondents). The loss of biological and non-biological NRs emerges as the most important ecological effect related to armed conflicts in and around the PNKB in the Lulingu sector.

8. Socio-economic effects of armed conflict on the park and the surrounding population in the Lulingu sector

Results on the socio-economic effects of armed conflict on the park and the surrounding population in the Lulingu area are also presented.

Table 15: Distribution of respondents on the socio-economic effects of armed conflict on the park and the surrounding population in the Lulingu sector.

Socio-economic effects of armed conflict on the park and the surrounding population	Frequency	%
Loss of revenue from tourism activities	55	42,3
Loss of work for park agents	24	18,4
Loss of benefits initiated by the park for the local population	51	39,2
Total	**130**	**100**

Source: our surveys

The results of this table show the socio-economic effects of armed conflict on the park and the local population as a result of human activities carried out by armed groups in the KBNP. The respondents identified three types of socio-economic effects of these activities undertaken by the armed groups on the KBNP and the surrounding population, namely the loss of revenue from tourism activities (42.3% of respondents), the loss of benefits initiated by the park for the surrounding population (39.2% of respondents), and finally the loss of work for park agents (18.4% of respondents) The loss of revenue from tourism activities and the loss of benefits initiated by the park for the local population emerged as the most important socioeconomic effects related to the armed conflicts in and around the KBNP in the Lulingu sector.

9. Mechanisms to address the presence of armed conflict actors in and around the KBNP in the Lulingu area

Results regarding mechanisms to address the presence of armed conflict actors in and around the KBNP in the Lulingu area are presented.

Table 16: Distribution of respondents on mechanisms to combat the presence of armed conflict actors in and around the KBNP in the Lulingu sector.

Mechanisms to address the presence of armed conflict actors in and around the park	Frequency	%
Strengthening park monitoring services through training and equipment	15	11,5
Restoration of the authority of the State through the defense, security and administrative services	67	51,5
Sensitization of armed groups on the importance of the park	26	20
Enforcement of legal and penal requirements for the protection of species of fauna and flora in protected areas	22	16,9
Total	**130**	**100**

Source: our research

This table presents the different types of mechanisms proposed by respondents to combat the presence of armed conflict actors in and around the KBNP in the Lulingu area. The four types of mechanisms proposed by respondents are listed here in order of importance as follows the restoration of state authority through the defense, security and administrative services (51.5% of respondents), sensitization of armed groups on the importance of the park (20% of respondents), the application of legal and penal prescriptions relating to the protection of species of fauna and flora in protected areas (16.9% of respondents), and finally the strengthening of the park surveillance services through training and equipment (11.5% of respondents)

In light of these results, it is clear that the restoration of state authority through the defense, security, and administrative services and the sensitization of armed groups on the importance of the park are the two mechanisms proposed by respondents to effectively combat the presence of armed conflict actors in and around the KBNP in the Lulingu sector.

3.3.3. Discussion of the results

This research had the following specific objectives:

i) Identify the reasons for the presence of armed conflict actors in and around the KBNP in the Lulingu area;

ii) List the effects of the presence of armed conflict actors on the KBNP in the Lulingu area in order of importance;

iii) Propose sensitization on the importance of the park as a mechanism that can contribute to fight against the presence in and around the PNKB in the sector of Lulingu of actors of armed conflicts.

The discussion of the results will focus on these three specific objectives as presented above.

Regarding the first objective of this work, the results clearly showed the presence of armed conflict actors or armed groups in and around the KBNP in the Lulingu sector, particularly the FARDC and the RM (60.7% of respondents): the weakness in the restoration of state authority in the area (60.7% of respondents), the ignorance of the importance of the park by these armed groups (23.8% of respondents) and finally the weakness of the park surveillance service (15.3% of respondents): The covetousness of the park's NRs (56.1% of respondents), war strategies (36.6% of respondents) and finally the protection of the park and its NRs (6.9% of respondents), while the respondents recognized 4 types of activities carried out by these armed groups in and around the park, namely Illegal extraction of minerals (48.4% of respondents), poaching of animals (36.9% of respondents), agro-pastoral exploitation (10% of respondents) and finally illegal exploitation of wood and PNFL (6.1% of respondents).

These results are also confirmed by BOYZIBU (Sd) by pointing out that 90% of the PNKB is occupied by armed gangs and only 10% is currently **controlled.** In his study conducted in a number of African PAs GUILLAUME (2009) attests to our findings by saying that at present, countries that are the scene of armed conflicts generally possess exceptional natural resources whose exploitation offers extraordinary economic profitability.

While BANNON and COLLIER (2003) and LA DOCUMENTATION FRANÇAISE (2004) affirm that the lack of organization of States and the control of a few leaders, often armed, over the production and distribution networks of raw materials constitute an extremely fertile ground for the development of inequalities, corruption and the anarchic exploitation of men and nature, giving a well-known example is the DRC, which possesses on its land and in its subsoil a remarkable deposit of highly coveted mineral, animal and vegetable raw materials. Indeed, the country "is endowed with an abundance of rare mineral resources from the northeast to the southeast of the country (coltan (colombium and tantalum), diamonds, gold, copper, cobalt, zinc, manganese, etc.), very rich forest and wildlife resources (gorillas, okapi, etc.) and vast fertile soils suitable for agriculture (coffee, tobacco, tea, etc.).

For the UNDP (2005), the causal link between the origin of armed conflicts and the presence of natural resources is very clear. This is also what emerges from the 2005 annual development report produced by the UNDP, in which it is noted that "Between 1990 and 2002, the world experienced at least 17 conflicts of this type in which the abundance of natural resources was the major factor. Diamonds in Angola and Sierra Leone, timber and diamonds in Liberia, gemstones in Afghanistan, and copper, gold, cobalt and timber in the Democratic Republic of Congo have all been at the center of civil conflict, or, in the case of the Democratic Republic of Congo, incursions supported by neighboring states. Other ethnic, territorial and historical causes complicate the understanding of the conflicts.

Regarding the effects of armed conflict on the KBNP in the Lulingu sector, the respondents recognized three types of ecological effects on the KBNP linked to the presence of armed groups in and around the park, namely the continued loss of fauna and flora species and other NR in the park (51.5% of the respondents), the loss of habitat for the park's fauna and flora (33% of the respondents), and the pollution of the park's ecosystem (15.5% of the respondents),5% of respondents), which directly associate three types of socio-economic effects, namely the loss of revenue from tourist activities (42.3% of respondents), the loss of benefits initiated by the park for the local population (39.2% of respondents), and finally the loss of work for park agents (18.4% of respondents).

Our results are consistent with those found in a paper by ANNE et al (2009) in which these authors state that conservation activities can be affected (negatively) by violent conflict because the use of armed forces between two or more parties can be detrimental to the status and well-being of the beneficiaries of conservation (people, animals, ecosystems), as well as to the ability of conservation actors to carry out their activities. This direction of influence can be both direct and indirect.

While SHAMBAUGH et al (2001) show that armed conflict can specifically impact conservation activities not only directly: conflict destroys habitats and kills animals, natural resources are over-exploited for both survival and profit, relief shelters and camps create new sources of pollution, park staff are threatened and often killed by armed groups, but also indirectly impact conservation activities: Conservation funding can dry up as nervous donors may withdraw their support and more immediate humanitarian needs may take precedence over environmental priorities.

In the same vein, SHUKU (Sd) in his study conducted in some PAs of the DRC has shown that armed conflicts have been at the root of the disruption of the ecological and economic balance of the forest ecosystems of the national parks and that this is manifested by facts such as the destruction of forests and biodiversity, the pollution of the ecosystems by explosive devices, the plundering of timber, mining and agro-pastoral resources and even the disappearance of certain rare animal and plant species and that this situation can be seen in all the PAs of the DRC where armed conflicts are taking place.

For HUGON (2001, 2003a, 2003b), from an economist's point of view, the "new" armed conflicts, which generally take place in poor countries, compromise economic growth and constitute an aggravating factor of underdevelopment and poverty.

It is in this context that the Ministry of Human Rights (1999) of the DRC has given the following economic situations for some national parks that are or have been subject to the presence of armed gangs: For Virunga Park: Deforestation: $34,104,000, Poaching: $139,338,000, Loss of income from tourism: $5,075,000, Other ecological damage: $21,291,000TOTAL: $199,808,600; For Kahuzi-Biega Park: Deforestation: $11,368,000, Poaching: $11,368,000, Lost tourism revenue: $2,450,000, Other ecological damage: $8,124,600TOTAL: $33,310,600; For Garamba Park: Deforestation, Poaching: $19,220,000, Lost tourism revenue: $2,500,000, Other ecological damage: $3,966,600, TOTAL: $25,686,600; Totals for the three parks: Deforestation: $45,472,000, Poaching: $169,926,000, Tourism shortfall: $10,025,000, Other ecological damage: $33,382,800, TOTAL: $258,805,800.

For the third specific objective of our work, the results led to four types of mechanisms proposed by the respondents to combat the presence of armed groups in and around the Lulingu sector of the PNKB the restoration of state authority through the defense, security and administrative services (51.5% of respondents), sensitization of armed groups on the importance of the park (20% of respondents), the application of legal and penal prescriptions relating to the protection of species of fauna and flora in protected areas (16.9% of respondents) and finally the strengthening of the park surveillance services through training and equipment (11.5% of respondents)

Thus, it is clear that the restoration of state authority is one of the mechanisms that can protect the PNKB from armed groups in the Lulingu area. This restoration of state authority requires the presence of strong administrative, legal and security institutions, i.e., full control of this entity by the territorial authorities and well-trained and disciplined police and military forces.

It is in this context that CHECHABO (2007) points out that in the DRC, the authorities in charge of the nature conservation sector, in this case the ICCN, do not have adequate material, financial and human resources to mitigate the threats of armed groups around and in the PAs. This lack of resources is exacerbated by the application of an inadequate monitoring system in these PAs. For this author, the lack of involvement of the state services that are supposed to assist the ICCN in the fight against poaching and their frequent involvement in poaching complicates the task of the ICCN. Unfortunately, the involvement of some agents of human organizations and security institutions in poaching and illegal exploitation of raw materials is also reported.

According to (GRIP, 1998, 2001; GEHRING, 2001; MOLLARD-BANNELIER, 2001), in the context of the use of legal prescriptions, jurists have also been interested in the correlations between armed conflict and the environment. Legal studies have focused on the implementation of and compliance with international conventions and treaties that can and must be applied during the course of hostilities, and for this purpose, three major legal instruments exist: the ENMOD Convention (1976), the Geneva Protocol I (1977), and the Guidelines of the International Committee of the Red Cross and Red Crescent (ICRC, 1996).

Our investigations have shown that the KBNP is besieged by armed groups in the Lulingu sector, and at the root of this presence is the weakness in the restoration of state authority in the area, leading to the exploitation of the park's natural resources by these groups. The ecological effects of this presence in the park are very diverse and have repercussions on the economic state of the park and on the socio-economic life of the local population.

Thus, the respondents proposed a number of mechanisms to combat the presence of these armed conflict actors in and around the KBNP in the Lulingu sector, including: the restoration of state authority through the defense, security, and administrative services; the sensitization of armed groups on the importance of the park; the application of legal and penal prescriptions relating to the protection of fauna and flora species in protected areas; and finally, the reinforcement of the park's surveillance services through training and equipment.

In view of the above, as researchers, we have the obligation to make a number of suggestions and recommendations in the following chapter to all levels so that each of them can take their responsibilities to help ensure better protection of the PNKB and its NR by improving the security situation in and around the park in the sector of Lulingu to make it safe from armed groups for its sustainability

SUGGESTIONS AND RECOMMENDATIONS

The consideration of the global environment in the context of sustainable development which is based on economic, social and environmental aspects is very crucial.

Since 1994, with the massive influx of Rwandan refugees, followed by untimely wars, several armed groups considered by the DRC authorities as "negative forces" have established themselves inside the PNKB. The park is used as a rear base from which attacks on the periphery are launched on a more or less regular basis. These groups are also heavily involved in illegal mineral exploitation and poaching activities. The damage caused by mining results in the destruction of fauna and flora, water pollution and the detour of river beds. Hence the proposal for a Monitoring Framework for awareness raising and conflict resolution activities.

Table 17: Monitoring framework for awareness and conflict resolution activities

INDICATOR	METHOD	FREQUENCY	RESPONSIBLE
NUMBER OF FOCUS GROUP MEETINGS IN COMMUNITIES	Coco Report	Quarterly	Coconut agents
NUMBER OF EMISSIONS RADIO TELEVISED AND BROADCAST	Surveys Polls	Quarterly	Communication
NUMBER OF WORKSHOPS, SEMINARS AND CONFERENCES HELD	Coconut report	Quarterly	Coconut agents
NUMBER OF DOCUMENTS PRODUCED, PRINTED AND DISTRIBUTED	Coconut report	Quarterly	Communication
NUMBER OF CONFLICT RESOLUTION INTERVENTIONS	Coconut report	Quarterly	Area Managers
NUMBER OF MEETINGS WITH TARGET GROUPS IN COMMUNITIES	Coconut report	Quarterly	Coconut agents

However, the DRC, facing armed conflicts in its eastern part, the PNKB, one of its national parks, has also been facing for several decades the presence of armed groups inside as well as on its periphery, who engage in human activities of a nature to compromise its overall ecological balance, This can compromise not only the ecological and tourist value of this park but also the well-being of the local population in terms of ecological and socio-economic services from which it benefits from this park.

In this context, both public and private actors must make efforts to contribute to the fight against the presence of armed groups in the KBNP in the Lulingu sector. However, the combination of efforts in synergy between actors would lead to effective and sustainable actions. This being the case, all actors involved in the field of nature conservation at all levels must be called upon to assume their full responsibilities. On our part, our contribution as a researcher in the field of the environment in general and nature conservation in particular is limited to suggesting and recommending the following:

To the international community of :

- Encourage states and national armed forces to protect the natural environment and/or PAs during armed conflict by taking appropriate measures;

- Facilitate the training of armed forces in an often neglected area of international humanitarian law, that of the protection of the natural environment;

- Prohibit the use of methods and means harmful to the natural environment and/ or PAs during armed conflicts in which only military objectives are attacked, but not the environment or PAs.

To the Congolese State of :

- Implement a study project and strategic action plan to correct and mitigate war-related impacts;

- Initiate good governance mechanisms throughout the national territory, including the Lulingu entity;

- Initiate good collaboration with neighboring countries and all other partners;

- To arouse the love of the fatherland and the national unity in the heads of the Congolese populations in order to break any impulse of war in our country;

- Promote the sound management of national resources and regulate their use;

- Strengthen the supervision of the populations on the national territory, including that of Lulingu, to enable them to preserve and conserve animal and plant species by creating, for example, certain economic activities that can make protein available (fish farming, livestock breeding, food canteen);

- To initiate sustainable mechanisms that will allow for the total demobilization of all young militiamen in all corners of the country, especially those in the territory of Shabunda;

- Strictly respect the existing forestry legislation;

- Create other animal species reserves;

- Train and inform national security and defense forces on the importance of PAs and the mechanisms for their protection, especially in times of war.

At the ICCN of :

- Enforce PA conservation and surveillance laws in peacetime and wartime;

- Train and retrain PA staff, rehabilitate and build infrastructure, and equip PAs to deal with untimely attacks by armed groups;

- Initiate socio-economic studies, environmental education, and allow local communities to integrate the management of PAs in their areas as local development initiatives;

- Consolidate scientific knowledge as a basis for making appropriate and adequate management decisions;

- To ensure a greater and more productive exploitation of the tourist potential of our natural heritage is also poorly exploited to create benefits for the country and the population;

- To further guarantee security for our biodiversity and to contribute to the consolidation of peace in the Central African region through cross-border and regional cooperation;

- Strengthen the partnership between other international and national institutions in order to guarantee the capacity building that ICCN needs to better assume its missions in the difficult situation our country is going through and in times of war;

To the PNKB of (from):

Strengthen the capacity of its personnel involved in securing the boundaries of the park assigned in the Lulingu sector through appropriate training sessions;

Sensitize the armed groups present in and around the park in the Lulingu sector and the local population on the importance of the park and its sources;

Initiate income-generating activities to help keep local militia youth busy and demobilize them.

At IUCN

To practice the polluter pays principle or to pay the ecotax, also called ecological tax;

That the industrialized countries of the planet finance the projects of conservation in Africa because it possesses a not insignificant quantity of the biodiversity.

GENERAL CONCLUSION

This work on "the effects of armed conflicts on protected areas in the DRC: the case of the Kahuzi-Biega National Park (KBNP) in the Lulingu sector" aimed to contribute to the protection of the KBNP and its natural resources at all times, even during periods of armed conflict, with a view to maintaining the overall ecological, socio-economic and even cultural value of the park. The research had the following specific objectives:

i) Identify the reasons for the presence of armed conflict actors in and around the KBNP in the Lulingu area;

ii) List the effects of the presence of armed conflict actors on the KBNP in the Lulingu area in order of importance;

iii) Propose sensitization on the importance of the park as a mechanism that can contribute to fight against the presence in and around the PNKB in the sector of Lulingu of actors of armed conflicts.

This work is structured in four chapters, except for the general introduction and the conclusion, of which the following are included

The first chapter presents the general introduction;

The second chapter gives the literature on the subject;

The third chapter discusses the survey, analysis and interpretation of the results;

The fourth chapter presents suggestions and recommendations.

In order to collect data and to get an idea of the reality on the ground, we conducted a survey among the population, which included several strata, including the PNKB agents in the Lulingu sector. This survey was completed by observations and free interviews with resource persons.

After the field research, the following results were obtained:

100% of respondents recognized the presence of armed conflict actors or armed groups in and around the KBNP in the Lulingu sector, particularly the FARDC and the MR (60.7% of respondents) and the basis for the presence of armed groups in the KBNP in the Lulingu sector or are weakness in the restoration of state authority in the area (60.7% of respondents), ignorance of the importance of the park by these armed groups (23.8% of respondents) and finally the weakness of the park surveillance service (15.3% of respondents);

There are three types of reasons why armed groups are currently present in the KBNP, namely the coveting of the park's natural resources for 56.1% of respondents, war strategies for 36.6% of respondents, and finally the protection of the park and its natural resources for 6.9% of respondents. From these results, it is clear that the actors of armed conflict are currently present in the KBNP in the Lulingu sector because of their desire to covet the park's NRs, while 100% of the respondents recognized that the armed groups present in and around the KBNP carry out human activities in the park, including 4 types of activities, namely the illegal extraction of minerals (48,4% of respondents), animal poaching (36.9% of respondents), agro-pastoral exploitation (10% of respondents), and finally the illegal exploitation of wood and PNFL (6.1% of respondents);

Three types of ecological effects on PNKB linked to the presence of these armed groups in and around the park were recognized by the respondents, namely the continued loss of species of fauna and flora and other NR in the park (51.5% of respondents), the loss of habitat for the park's fauna and flora (33% of respondents), and finally the pollution of the park's ecosystem (15,5% of respondents), which directly associate three types of socio-economic effects, namely the loss of revenue from tourist activities (42.3% of respondents), the loss of benefits initiated by the park for the local population (39.2% of respondents), and finally the loss of work for park agents (18.4% of respondents);

Four types of mechanisms were proposed by respondents to combat the presence of armed conflict actors in and around the Lulingu sector of the KBNP the restoration of state authority through the defense, security, and administrative services (51.5% of respondents), sensitization of armed groups on the importance of the park (20% of respondents), the application of legal and penal prescriptions relating to the protection of species of fauna and flora in protected areas (16.9% of respondents), and finally, the strengthening of the park's surveillance services through training and equipment (11.5% of respondents).

BIBLIOGRAPHIC REFERENCES

SCIENTIFIC BOOKS AND ARTICLES

PATRICK T., 2009. Manuel de gestion des aires protégées d'Afrique francophone. Awely, Paris, pp.1215, "hal-00669157";

RONGERE P., 1971. Méthodes des sciences, Paris, Dalloz;

SOURNIA G et al., 1998. Les aires protégées d'Afrique francophone, Paris, Ed. Jean-Pierre de MONZA, 272p;

AL-HAMANDOU D. and MICHEL-ANDRE B., 2016.Armed conflicts and the environment: Framework, modalities, methods and role of Environmental Assessment. Développement durable et territoires, France.;

ANNE H., ALEC C., ROBERT C., ROBERT M. AND RICHARD M., 2009. Toward conflict-sensitive conservation: a practitioner's manual. International Institute for Sustainable Development (IISD). 75p;

ANONYMOUS (SD) The covetousness of the DRC's wealth and its consequences on the protection of the national environment;

BANNON I. and COLLIER P., 2003. Natural resources and violent conflicts. Options and Actions. Washington DC, The World Bank, 429 p.;

BOYZIBU E., Sd. Impacts of a decade of war on natural resources in protected areas in DRC, ICCN;

BUSANE R. W., 2006. The Management of Protected Areas in South Kivu: Practices and Conflict;

CHARLINE B. and CHARLES H., Sd. "TOWARDS CONFLICT-SENSITIVE PROGRAMMING" Search for Common Ground (SFCG) DRC Training Module;

FISCHER, 1993. The vegetation of Kahuzi-Biega National Park (South Kivu/Zaire), Born, 93.p;

GEHRING R., 2001. La protection de l'environnement en période de conflit armé: Que peut ou pourrait apporter la Cour Pénale Internationale, Université de Lausanne, Février, 18 p.;

HUGON P. 2001, 2003a, 2003b. The Economics of Conflict in Africa; The Economics of Conflict; Armed Conflict in Africa: Myth and Limits of Economic Analysis;

HUMAN RIGHTS WATCH (HRW), 2009. "You Are Being Punished": Attacks on Civilians in Eastern Congo, December 13, 2009;

HEDBERG O., 1957. Afroalpine vascular plants/À taxonomic revision. Symb. Bot. Upsal, 15: 1-411;

MUHIGWA B. J-B., BAKONGO M. G., JAAP S. & LETICIA P. C., 2007. Study of the development of a buffer zone of the PNKB in the community sector MUDAKA-IZEGE via an intense reforestation, PBF-GTZ-PNKB;

MÜHLENBERG M., SLOWIK J. & STEINHAUER-BURKART B., 1994. Kahuzi-Biega National Park. Brochure published by the Zairian-German project IZCN/GTZ, Bukavu. Integrated Nature Conservation: 1-52 p.;

RUNGE J., 1992. Geomorphological observation concerning palaeo environmental conditions in eastern Zaire. Z. Geomorph. N.E., supplement 91: 109-122;

SHAMBAUGH, J., OGLETHORPE J. and HAM R., 2001.The Trampled Grass: Mitigating the Impacts of Armed Conflict on the Environment, WWF Biodiversity Support Program,Washington, DC;

POURTIER R., 2009. "Le Kivu dans la guerre: acteurs et enjeux," Echo Géo;

VANSINA J. 1990. Paths in the Rainforests - Towards a History of political tradition in equatorial Africa.London: James Currey;

WHITE L. and EDWARD A., 2001. Conservation in the African Rainforest, research method, France, 455.p.

DICTIONARIES

RAMADE F., 2002. Dictionnaire encyclopédique de l'écologie et des sciences de l'environnement, Dunod, Paris, 1175.p;

ERRI, P., 1988. Dictionary of the International Law of Armed Conflict, ICRC, Geneva, 36p.

THESES AND DISSERTATIONS

CHECHABO B. B., 2007. Environmental Impact of Population Displacement in Situations of Armed Conflict: The Case of Refugees in the Eastern Democratic

Republic of the Congo Master pro (M2) in International and Comparative Law. Limoges/Faculty of Law and Economic Sciences;

EMMANNUELLE G., 2014. Biodiversity conservation strategies and the process of prioritizing actions in armed conflict zones in Central Africa. Master's thesis. University of Sherbrooke. Quebec. Canada;

GUILLAUME B., 2009. Les évaluations des impacts sur l'environnement en période de conflits armés, Université de Sherbrooke, for the double master's degree in environment and master's degree in engineering and management in environment and sustainable development;

KATOTO M. W., 2012. Anthropogenic pressures on animal and plant biodiversity in the Kahuzi-Biega National Park (KBNP): Bitale-Itebero axis (DR Congo). Master's thesis, University of Burundi, 115p;

MOLLARD-BANNELIER K., 2000. La protection de l'environnement en temps de conflit armé, Université de Paris I (thèse de Doctorat), Paris Pedone, 542 p;

SHUKU O. N. (Sd). The impact of armed conflict on forest ecosystems and national parks in the DRC 5p.

OTHER REPORTS

GRIP, 2001. Report of the Group of Experts on the Illegal Exploitation of Natural Resources and Other Forms of Wealth in the Democratic Republic of Congo (DRC), Brussels, April, 62 p;

HART J.M., CARBO F., AMSINI F. GROSSMANN and C. KIBAMBE, 2007. Kahuzi-Biega National Park, lowland sector: Preliminary inventory of large fauna. Unpublished report: Inventory and Monitoring Unit of ICCN Report No. 7, 49p;

HUMAN RIGHTS, 2009. Report: Democratic Republic of the Congo, Washington: U.S. Department of State, March 11, 2009;

ICC, 2016. Report on the state of conservation of DRC properties inscribed on the list of World Heritage in Danger fiscal year 2015, Branch 61p;

ICCN, 2007. GEF-BM Project. Social Impact Assessment. Final report;

ICCN, 2010. Parks and Reserves of the Democratic Republic of Congo. Evaluation of the effectiveness of protected area management;

ICCN-PNKB, 2009. General Management Plan (GMP), 2009-2019;

KIPPENBERG J. et al, 2009. Violent soldiers, commanders turn a blind eye. Sexual Violence and Military Reform in the Democratic Republic of Congo, New York: Human Rights Watch;

General management plan of the PNKB 2009-2019;

UNDP, 2005. Human Development Report 2005 (chap. 5, p. 161193), Paris, Economica;

UWE K. and TERESE H., 2006. Rapport de Mission de suivi réactif au Parc national de Kahuzi-Biega République démocratique du Congo (RDC). German Technical Cooperation (GTZ).

LA DOCUMENTATION FRANÇAISE, 2004 and AUTRE LOI.

The plundering of the natural resources of the DRC;

NGUIFFO S., 2003. The Forestry Law and the Marginalization of Pygmy Populations, CENTER FOR ENVIRONMENT AND DEVELOPMENT (CED).

<u>WEBSITES</u>

http://echogeo.revues.org/10793;

<u>http://wilkipedia/wiki/, lord's resistance army</u>, accessed 04/03/2017;

<u>http://www.ladocumentationfrancaise.fr/dossiers/conflit-grands-lacs/pillage</u> ressourcesnaturelles-RDC.shtml (Page consulted on June 30, 2009);

<u>http://www.unmultimedia.org/radio/english/2012/06/un-human-rights-chief-fears-more rapes-killings-in-congo-by-m23/,</u> accessed 23/03/2017.

TABLE OF CONTENTS

Printed by Books on Demand GmbH, Norderstedt / Germany